BUG BITES

INSECTS HUNTING
INSECTS... AND MORE

DIANE
SWANSON

Photographs and illustrations
from the
Royal British Columbia Museum

D1370468

Whitecap Books
Vancouver / Toronto

Edited by Elaine Jones
Proofread by Elizabeth McLean
Cover and interior design by Carbon Media
Cover photographs by the Royal British Columbia Museum
except photograph on left which is by Steve Marshall

Printed and bound in Canada.

Canadian Cataloguing in Publication Data

Swanson, Diane, 1944 –
Bug bites

Includes index.
ISBN 1-55110-532-2

1. Insects—North America—Juvenile literature. I. Title
QL467.2.S96 1997   j595.7'097   C96-910732-3

THE CANADA COUNCIL | LE CONSEIL DES ARTS
FOR THE ARTS | DU CANADA
SINCE 1957 | DEPUIS 1957

We acknowledge the support of the Canada Council for the Arts
for our publishing program.

We acknowledge the support of the Cultural Services Branch of the
Government of British Columbia in making this publication possible.

CONTENTS

Small is big in the insect world. Most insects are less than 6 millimetres (about .25 inches) long, which means they don't need much space. They can live many places in huge numbers — even in your backyard.

Besides being small, insects adapt well. They can live all over the world in all kinds of homes, from sea coasts to mountain tops. Some live in snow; others, in deep mines. Some live on the surface of oceans. But then, insects have had over 400 million years of experience living on this planet. That's much more experience than people have had.

The total number of different kinds of insects is greater than the total of all other kinds of animals put together. Scientists have discovered and named about one million kinds of insects, but there are likely millions more.

The populations of insects are also huge. About 2 million termites can share one mound. About 750 million ladybugs can spend winter in one clump. Billions of soil insects can live in one meadow the size of a football field.

Insects are awesome reproducers. Many lay masses of eggs that produce huge numbers of insects. The eggs of one kind of wasp can produce — not 3 (triplets), not 5 (quintuplets), but — 1000 identical wasps EACH. (What would you call 1000 identical wasps — "milliplets"?)

Ladybugs that gather to rest through cold
months sometimes blanket huge areas.

GOOD BUGS, BAD BUGS?

When people ask if an insect is "good" or "bad," they usually mean, "Will it attack me?" or "Will it harm my garden?…my trees?…my crops?"

Of all the insects in the world, very few kinds bother anyone in any way. Even wasps, which scare a lot of people, usually just sting other insects and spiders — food for their young. Wasps that sting people are likely trying to defend themselves or their home.

Yet many people fear insects. Part of the reason is because most insects are very odd-looking to us — even scary. And if they are not walking around on six hairy legs, they are zooming around on one or two pairs of wings. For years, movie producers have taken ideas from insects to make hair-raising horror films.

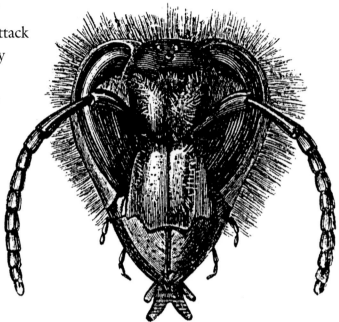

Up close, the awesome head of a hornet can be a frightening sight.

Even nice-looking insects, such as ladybugs, can frighten people. During the long, hot summer of 1976, there were many, many ladybugs in Britain, but there was very little moisture. The insects were attracted to human sweat, settling on skin to sip. But many people thought that the ladybugs were attacking.

Like other kinds of animals, insects are not "good" or bad." They are just doing whatever insects do. And it's a good thing, too. Their role in nature helps keep the world going around. Insects of many kinds are important food for fish, birds, and other animals. Many insects pollinate plants of all kinds. Others eat dead plants and animals, recycling nutrients and cleaning up the world.

INSECTS ON THE HUNT

It's true that some kinds of insects feed on some of the same plants that people do. But other kinds feed on many of these plant-feeding insects. They all help nature keep a balance, which helps people, too.

Insects that hunt to feed themselves or their young are some of the most fascinating animals anywhere. They are among the smallest, but most aggressive, hunters — the lions and tigers of the insect world. And they use astonishing techniques to capture their prey.

Ant lions dig pits in the sand, then wait in hiding to grab their victims. Robber flies pounce on insects in flight, inject them with poison, and drain them dry. Tiger beetles race out of hiding to snatch prey in powerful mouths. Dragonflies scoop flying insects in baskets formed by spines on their legs. And giant water bugs — likely the most powerful of all insect hunters — use poison to attack small snakes, fish, and insects.

Incredible but real-life dramas, such as these, occur often in almost every field, forest, pond, and yard. And the stars of these dramas are the insects — the super mini-beasts of the world.

PEOPLE ON THE HUNT

Insects might frighten people, but people might frighten insects even more — if the insects were able to think about it. Besides spraying chemicals and setting traps to kill billions of insects, many of the world's people have eaten them. For centuries, families in Asia, Africa, and Latin America have depended on insects for food.

Even today, many people — especially those who can't afford much chicken, beef, or fish — eat insects for protein, vitamins, and minerals. Outside of big cities, Mexicans regularly eat more than 100 different kinds of insects — usually fried or roasted. In Asia, people enjoy giant water bugs, steamed, fried, or ground up in sauce. They also eat the bugs' eggs.

Children in Indonesia wave sap-covered poles in rice fields to capture dragonflies. When the insects perch on these poles, they get stuck. All but the dragonflies' wings is fried in a spicy oil until crisp.

CHAPTER 2

The fiercest dragon of field and pond is the fastest insect that flies. The dragonfly hunts in bursts of speed up to 55 kilometres (about 35 miles) an hour. It scoops up insects midair in a net formed by its spine-covered legs. In just hours, it catches and crunches its own weight in food.

DASHING DRAGONS

Having the keenest sight in the insect world helps the dragonfly target its prey. Two enormous eyes see almost all directions at once. Each eye has up to 30 000 lenses: big ones for looking distances, small ones for checking things up close. It takes most of the dragonfly's brain to make sense of everything these amazing eyes see.

Sometimes the dragonfly eats its prey on the go; sometimes it waits until landing. Long front legs hold the prey firmly in place while a strong, toothy mouth grinds it up. One part of the mouth works like a sharp fork, turning the prey over.

Small insects, such as flies, make common dragonfly food. But some dragonflies also gobble butterflies and bees. Now and then, large dragons — with wings that spread up to 130 millimetres (5 inches) — even attack hummingbirds.

After feeding, the dragonfly sweeps its eyes clean, using a set of spines on its front legs. It's ready to hunt again.

A small, orange dragonfly is helpless in the tight grip of this larger dragonfly.

COLORFUL DRAGONS, COLORFUL DAMSELS

Thousands of kinds of dragonflies live around the world. Their wings — and their bodies — come in red, orange, yellow, green, blue, white, and black. On some kinds, wing patches flash scarlet or gold in the sun.

There are also many, many different damselflies — slim, slower-flying insects that are commonly called dragonflies. A damselfly looks like a dragonfly and acts like a dragonfly. But the damselfly is usually smaller, and it holds its wings together at rest — unlike the dragonfly. Still, the two are very close relatives. They share most of the same behavior.

A damselfly is stuck in goo from an insect-eating plant called a sundew.
When the leaves fold around the damselfly, the sundew will digest it.

ACES IN THE AIR

Dragonflies are flight masters. Sixteen muscles power each of their four wings, which can beat more than 40 times a second. Unlike other insects, dragonflies beat their front and back wings independently. When one pair goes down, the other pair goes up. The wings can bend and twist, too.

All this makes dragonflies amazingly speedy and nimble in the air. Besides zooming forward, they fly up, down, and sideways. For short distances, they can fly backward. They even hover midair. To dodge predators, such as birds and frogs, dragonflies take off fast and make sudden turns. Even if they damage or lose one pair of wings, they can still fly.

Researchers have noticed that dragonflies in strong wind twist their wings on the downbeat. That makes the air whisk across the tops of their wings, helping the dragonfly rise. People who design planes and helicopters get ideas from watching this insect.

When it flies, the dragonfly depends on more than its wings. It uses its head, holding it level with the ground. That helps the dragonfly keep its balance as it slants its body one way or another. And when it needs to put on the brakes, it lowers its back end and legs, slowing right down.

Dragonflies usually fly low over water and land, but they can — and do — fly high at times. One researcher in a plane spotted dragonflies 2100 metres (7000 feet) above ground.

Most dragonflies are daytime fliers, especially active in bright sunshine. In fact, many of them rest if the day turns cloudy, heading for the same plant or log each time.

The dragonfly uses its strong sense of direction to return to its home pond.

DOCTORS, STINGERS, AND DARNING NEEDLES

The fast flight and fierce face of the dragonfly have frightened people for centuries. No wonder this huge insect has earned several nicknames and inspired tall tales.

As the "devil's darning needle," it was said to sew up lips, nostrils, ears, or eyelids of children who misbehave. Some people believed it searched beds at night for uncovered fingers or toes — and sewed them together.

Dragonflies were also believed to be "snake doctors." They guarded water snakes and helped them find food. People feared a snake attack if they killed a dragonfly.

Thinking dragonflies could sting, some people called them "horse stingers." The insects were thought to sting bad children but help good children find great spots to catch fish.

THE DRAGON'S EMPIRE

Male dragonflies claim parts of ponds and stretches of streams as hunting territories. Big dragonflies often have big territories, some spanning 100 metres (about 330 feet) along the water's edge.

The dragonfly defends his empire against invaders. He may threaten first or head straight into battle. He may bolt upward and strike an invading dragonfly from below. Or the two may charge head on, their wings clashing noisily. They may even try to bite each other's legs.

Dragonfly battles sometimes end in injury or death. But most often, the invader simply leaves the territory. The defender may chase after him, just to make sure he's gone. So determined is the dragonfly to defend his empire that a large one may drive off a bird, such as a barn swallow.

Dragonflies hunt for more than food in their territories. They patrol for mates. When a male chases a female, he is beginning a mating ritual that is unique. He grabs the female's head with strong claspers at the end of his body. She curls her body under his to fertilize her eggs. For a few seconds, they fly together, joined in an oddly shaped loop. If dragonflies have longer-lasting unions, they perch to mate.

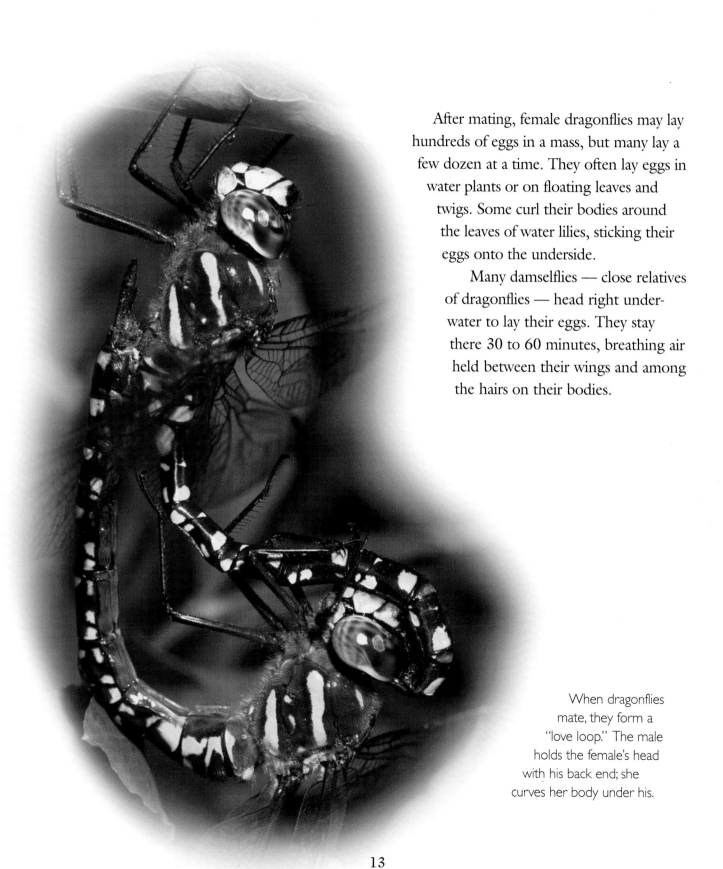

After mating, female dragonflies may lay hundreds of eggs in a mass, but many lay a few dozen at a time. They often lay eggs in water plants or on floating leaves and twigs. Some curl their bodies around the leaves of water lilies, sticking their eggs onto the underside.

Many damselflies — close relatives of dragonflies — head right underwater to lay their eggs. They stay there 30 to 60 minutes, breathing air held between their wings and among the hairs on their bodies.

When dragonflies mate, they form a "love loop." The male holds the female's head with his back end; she curves her body under his.

GIANT OLDIES

Of all the insects on Earth, the dragonfly is among the very oldest, and the first to fly. Remains of one dragonfly were found in France in a rock that is 300 million years old.

Except for details in its wing, this ancient dragonfly looked much like modern ones. But its wingspread was at least six times greater. The early insect had a body 300 millimetres (12 inches) long and a wingspread of 750 millimetres (30 inches). That's a body longer than a loose-leaf page and a wingspread nearly four times wider.

WATER DRAGONS

Dragonflies may take less than four weeks to hatch from eggs but up to five years to become adults. During that time, the young dragonfly — called a larva — lives underwater. Looking nothing like the adult, it goes through 10 to 15 stages, shedding each time it outgrows its outer coat.

The dragonfly larva is able to breathe in water by using gills inside its back end. It takes in oxygen from the water that it pumps past these gills. Sometimes the larva also uses this water to escape danger, firing it out to jet itself forward. In seconds, the larva can flee the jaws of turtles, fish, frogs, and snakes.

The damselfly larva can't do that. Its leaflike gills are outside its body, so it doesn't draw water in. But the swishing movement of the gills helps the damselfly swim.

This old drawing shows stages in the life of a dragonfly: water larvae (notice the graspers used to snatch prey); a new adult pulling out of its old coat on a leaf; and a fully developed adult, flying.

A dragonfly larva hunts as fiercely as an adult. Some kinds creep among water plants, searching for prey. Others hide in mud and rubble, waiting for prey to come close.

The larva uses a unique trap to capture prey. It snaps out a lower lip, which is almost half as long as its body. Spiny, jawlike graspers at the tip of the lip snatch up insects, tadpoles, and small fish. Then the lip draws the prey into the larva's real mouth.

When the lip isn't in use, the larva folds it up. Part of it forms a mask over the larva's face; part lies beneath the head. But if more prey strays within range, the lip shoots out again.

One day, one year, the larva enters the most exciting stage of its life. It climbs up a plant so that it's just a few centimetres (an inch or so) above the water. There, it goes through one final stage, splitting its coat and emerging as an adult dragonfly. Its breathing system must adjust from working in water to working in air. Once out of the watery world of the larva, there's no going back.

At first, the adult dragonfly is weak and colorless. Several days may pass before it gains its rich shades. Then it takes to the air — an instant master of flight. Ahead lie up to five splendid weeks of hunting as the dashing dragon it was meant to be.

THE MOVING MASSES

As young adults, dragonflies often spend time feeding away from the water. Sometimes they travel in crowds — occasionally, huge masses — but no one is sure why. Imagine how they must have amazed people in each of the following places.

- **Germany:** In 1862, more than 2 billion dragonflies swarmed in one flight.

- **Sweden:** In 1883, a swarm of dragonflies took several days to fly over a single town.

- **Indian Ocean:** In 1896, a throng of dragonflies called globe-skimmers landed on a ship 1450 kilometres (900 miles) off Australia.

- **Canada:** In 1915, thousands of dragonflies perched on a telephone line, all facing the same way.

- **Egypt:** In 1923, dragonflies swarmed so thickly along one coast that they darkened the sky.

CHAPTER 3

Robbers may demand: "Your money or your life!" But robber flies offer no choice. They steal the life of their prey, attacking almost any insect they spot.

HIGHWAY ROBBERS

Named after last century's highway robbers — who chased travelers on horseback — speedy robber flies nab prey on the go. Many strike insects in flight; some pounce on prey at rest.

Watching for insects, robber flies may crouch on a fence, a tree, or a blade of grass. They turn their heads from side to side, even stretching up on their legs to scan for prey. Some kinds search by patrolling a particular stretch of field or by flying around exploring.

When they spot a target, these robbers use their long legs to grab it. In flight, they overlap the tough bristles on their four front legs to make a trap and scoop up insects. Sometimes, they also use their stronger back legs to hold wriggling prey very tightly.

The robbers spear their catch with a strong, beaklike mouth. They inject a poison that kills it at once, dissolving its insides. Then they drink it dry — in midair or on a perch.

Never fussy eaters, the robbers feed on whatever is around, including other flies. They are so tough they even attack big beetles, grasshoppers, and other insects that are larger and heavier than themselves. Some kinds also attack stinging insects, such as wasps and bees.

This whiskery robber fly looks almost comical, but as a hunter, it's ferocious.

Nabbing a butterfly is easy for a robber fly. It often attacks insects bigger than itself.

ROBBERS SMALL, ROBBERS TALL

Thousands of kinds of robber flies live around the world. They hunt in sunny fields, meadows, and gardens. Many prefer dry, sandy spots.

They are generally gray and hairy with big heads and bulgy eyes. Compared to the size of other kinds of flies, robbers are medium to large. Most are between 5 millimetres (.2 inches) and 30 millimetres (1 inch) long. But the biggest robber fly — a South American kind — is the biggest fly of all. Its length and its wingspan are about 65 millimetres (2.5 inches).

THE COPYCAT ROBBERS

It's hard to tell some kinds of robber flies from bumblebees. Both are big and fat. Both are covered with thick hair — yellow and black. Both buzz loudly as they fly around.

Looking and sounding like bees helps protect these robber flies. That's because many insect-eating animals leave stinging insects alone.

Even some small kinds of robber flies protect themselves by being copycats. Covered with gray and yellow hair, they look just like little leaf-cutter bees. And like these bees, they buzz around flowers as they hunt for prey.

COOL TOOL KIT

Robber flies have what they need to be successful. Large, strong wings power them rapidly through the air. Millions of years ago, the robbers' ancestors had four wings. But over time, the two back wings changed into balancers that look like tiny stems with knobs. Today they help the robber control where and how it flies.

Two huge eyes, each having thousands of lenses, give the robbers good sight. They use it especially to spot moving prey and to figure out how far away it is. Some kinds of robber flies seem to have extra good sight. From the air, they can spot a spider that blends in with the grass — even if the spider is barely moving.

Many kinds of robbers have beards and moustaches made of strong, prickly bristles. They help protect the flies from attacks by predators. They also protect the flies' big eyes from the struggling insects that robbers stab.

A bumblebee? Check again. This kind of robber fly mimics the bee to fool its enemies.

THE MAKING OF A ROBBER

Most female robbers simply scatter their eggs over the ground. Some use two long bristles at their back end to sweep dirt over the eggs. Some use spiny, platelike parts to push the soil aside, lay their eggs, then let the soil fall back over them. Other female robbers lay eggs on plants. Some leave masses of them on dead stems. Some cut into living stems and stick their eggs inside.

When a young robber — a larva — hatches from its egg, it looks like a tiny tube. It worms its way into the ground or into rotting wood where its body won't easily dry out. There it may feed on the larvae of insects such as beetles. It may also eat rotting plants.

After the larva has grown bigger and stored a lot of food in its body, it stops feeding. It even stops moving. Quiet and still, it starts to change into an adult. Its coat swells and bulges as the head, legs, and wings of the robber fly form underneath. Strong spines and thorns appear on its coat.

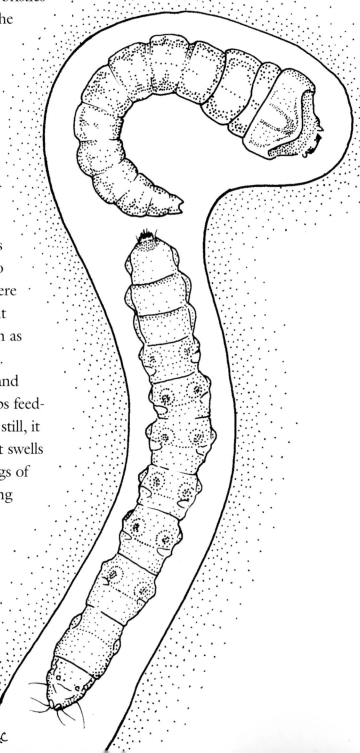

Ready to attack, a robber fly larva enters the wood burrow of a beetle larva.

Then, just as the robber is about to burst out, the larva heads toward the surface. It uses its spines and thorns to break its way through the ground. The old coat cracks, and out steps an all-new, all-tough highway robber, ready to steal a life.

As fierce a hunter as it is, this full-grown robber fly is being attacked by a tiny, red mite.

TRAVELING ROBBERS TRAPPED

A sea divides Alaska and Russia. But a wide bridge of land joined them once. Animals, including insects, could have crossed that bridge to get from one place to the other.

A robber fly that hunts in Alaska and Yukon is more closely related to robbers in Russia than to robbers south of Alaska and Yukon. Scientists think the robber fly may have crossed the land bridge about 20 000 years ago. Then the sea covered the bridge, trapping the robber in Alaska and Yukon.

CHAPTER 4

Killers with mouths like prickly needles prowl land and water around the world. These killers are bugs — true bugs.

KILLER BUGS

Some people call all insects "bugs," but true bugs are a special set of insects known for flat, leathery wings and a triangular plate behind their heads. They also have mouthparts, called beaks, that are built for piercing and sucking.

Of the thousands of different kinds of bugs world-wide, many feed on plants. Others are deadly hunters. They use their beaks to stab prey, injecting saliva that paralyzes it and turns its innards to liquid. Then the bugs suck up the juices.

Casting a dark, threatening shadow, an assassin bug quietly awaits its prey.

AWESOME ASSASSINS

Like the killer it's named after, an assassin bug hunts for its victims. This true bug creeps close, raises its beak like a dagger, and strikes. Muscular front legs grip the prey tightly until it struggles no more. Then the assassin feeds and slips its beak back into a groove between its legs.

The hairy front legs of some kinds of assassins ooze drops of thick, sticky liquid. It attracts small insects, such as fruit flies. When they try to feed on the liquid, they get stuck, and assassin bugs feed on them.

Like other bugs, assassins lay their eggs singly or in clusters. They often glue the eggs onto plants or fences to hatch. When the young bugs start hatching, they use special spines, called egg bursters, to break out. Then they leave the bursters behind, still stuck in the empty shells.

Except for size, newly hatched assassin bugs look much like the adults. They hunt the same way, too. But some of the young bugs stick bits of leaves and twigs on their backs so they blend in with the ground. Others glue bits of termites and termites' nests on their backs. That way, living termites don't notice the assassins that sneak up to their nest to hunt.

A close side view of the head of an assassin bug reveals a huge eye and needlelike beak.

ASSASSIN FASHION

Most assassin bugs are rather plain-looking, but some are quite striking. One kind has a raised, wheel-like crest on its back. Others are fringed with red or spotted with orange.

But one assassin bug can only be seen when it moves. Its sticky body and legs become covered with dust and lint, making it ghostlike. Unlike most assassins that hunt in fields, this one hunts inside houses.

Another assassin, called the thread-legged bug, also heads indoors, especially into sheds and barns. It is long and very skinny, with legs like threads.

Ready to nab almost any insect, this ambush bug hides by blending in with yellow flowers.

ARTFUL AMBUSHERS

Slow-moving ambush bugs hide and wait for prey to come near. That's how these true bugs got their name. Their usual greenish-yellow color and odd petal-like shapes help them blend in with the wildflowers they perch on. Some kinds look more like harmless ants, which fools insects into landing close by.

Although they are only up to 12 millimetres (.5 inches) long, ambush bugs prey on insects many times bigger. They attack butterflies, bees, and wasps. Strong, teeth-lined front legs help to hold the prey still.

WATER GIANTS

The biggest of the true bugs spends most of its life in the water. Called the giant water bug, it grows up to 75 millimetres (3 inches) long and hunts prey that's even bigger.

Some giants grab salamanders, fish, and small snakes, as well as tadpoles, snails, and insects.

Lurking among weeds in ponds and creeks, the flat, brownish giant often looks like a dead leaf. It takes cover until it spots a likely meal. Then it charges. It nabs its prey with mighty front legs and bites down hard.

Flattened back legs power the swimming of the giant water bug. It is well suited to living underwater. But it must surface to breathe now and then. Poking two short tubes just out of the water, it draws air in through its back end.

Pick up a giant water bug, and it may play dead — for as long as 15 minutes. Or it may fire a smelly liquid from its back end, just to drive you away. But sometimes, a giant leaves the water on its own. It may fly to a new pond or creek, searching for better water. Or it may just leave the water to fly at night.

Giants are so attracted to bright lights that they sometimes gather around street lamps. That's why some people call them electric light bugs.

A giant water bug stops hunting to take a breath — through its back end.

LIFE WITH FATHER

Doing push-ups underwater is one way male giant water bugs try to attract females. If they are successful, they mate. Then the females lay their barrel-shaped eggs, usually on plants above water. But some kinds of giant water bugs glue them to the backs of the males.

Laying up to 100 eggs in rows, the female covers the male so thickly that he can't fly. Even his swimming slows down under the load. For 10 days or so, the male carries the eggs wherever he goes. He protects them from predators and, again and again, takes them to the surface for air. He rocks back and forth. And he uses his hind legs to feel the eggs, checking all the rows from front to back.

So good is life with father that almost all the eggs hatch. Then the new little giants take off on their own.

DEADLY SCORPIONS

Don't let its slow walking and poor swimming fool you. The waterscorpion is a deadly killer. Clinging to under-water plants, it hangs head down with front legs raised. It's very still — but very ready.

When an insect or a small fish swims close, it may not even notice the waterscorpion. Then snap! Legs close tightly on the prey, and the scorpion spears it.

As it feeds, the waterscorpion may turn its prey over end to end, holding it with two legs or just one. It may poke its beak into several different parts to drain the prey.

Waterscorpions hang around — usually upside down — waiting for insects and tiny fish to swim by.

Most waterscorpions are sticklike — skinny and as long as 50 millimetres (2 inches). Others are leaflike — wide and only 12 millimetres (.5 inches) long. They get their name from the "snorkel" at their back ends, which some people think looks like the tail of a scorpion.

The snorkel is a long breathing tube made of two stiff, grooved threads hooked together. The waterscorpion pokes the end of the tube out of the water and draws in air.

Like many other bugs, the waterscorpion can fly. But it rarely does, except in emergencies. If its home pond dried up, it would fly away to find another place — even if it had to travel quite far.

BACKSWIMMING BITERS

True bugs called backswimmers attack other small insects, tadpoles, little fish, and more. Most backswimmers use their first four legs to clutch prey. Some kinds form a cage with these legs, using long bristles to hold their victims.

As you can guess, backswimmers swim on their backs. That works well because their backs are shaped like boats with keels. By pulling their long, strong back legs through the water, these bugs "row" well. Fringes of stiff hair make the legs broader and better for swift swimming. That helps the bugs zoom in on prey and scoot away from danger.

Backswimmers can also swim on their fronts. Researchers put some of these bugs in a tank lit only from the bottom. Attracted to the light, they swam toward it, but first, they flipped over — right side up. Keeping their pale, pearl-colored backs turned away from light helps protect them from predators. Their exposed dark undersides are far harder to see.

The backswimmer spends a lot of time floating near the surface of ponds. Its body is not as heavy as water, so floating is easy. What's hard is staying deeper. The bug has to cling to a plant or rock that is submerged.

Underwater, the backswimmer breathes from an air supply carried beneath its wings and in hairy grooves on its underside. It can tote enough air to breathe for up to six hours. When it runs low, the bug rises quickly, sticking its back end out to draw in air.

At the surface, a back-swimmer may turn a front flip and fly off — right side up. It can fly well. It can even fly far. Sometimes it joins crowds of other backswimmers drawn to bright lights at night.

During winter, some kinds of backswimmers sleep very deeply. Others still keep busy. In fact, you may spot a backswimmer walking upside down beneath the ice of a frozen pond. There, it may look more like a comic than a killer bug.

Up to 12 millimetres (.5 inches) long, the backswimmer uses its strong back legs to swim quickly.

CHAPTER 5

All day long a mantid waits with front legs raised and ready. Totally still or slowly swaying, it waits to catch a meal. Insects that land nearby may never notice the mantid. It looks like part of the plant that it's on. And it snatches insects so fast they barely see it move.

Using hundreds of lenses in two big eyes, the mantid is always watching for prey. Its eyes see well and adjust quickly to light. If the day is bright, they are tan or green; if it's dim, they turn dark. Changes to the eyes make it easier for the mantid to see.

Few insects can turn their heads up, down, and sideways. But a mantid can. It can also tilt its head and turn it to look back. All that movement helps it find prey.

And when it does, the mantid strikes in a split second — usually with perfect aim. With outstretched front legs, it seizes its prey. Strong muscles tighten the grip, and sharp spines pierce the prey, holding it firmly in place.

When the mantid chomps down, it slices easily through the tough outer coat of most insects. It nibbles its prey from end to end, feeding like someone eating corn on the cob. Soon all that's left are the wings and coat.

The mantid eats a lot. Even if it's stuffed with food, it will attack any prey that lands close by. It feasts on butterflies, aphids, moths, grasshoppers, flies, bees, dragonflies — even other mantids. Big mantids may also grab and gobble small birds, frogs, and lizards. After they feed, they nibble their front legs clean.

Perfectly still stands the mantid — until it strikes. Then its front legs move so fast you can't see them.

MANTIDS OF MANY COATS

Many parts of the world are home to mantids, and there are hundreds of different kinds. Some are as short as 10 millimetres (about .5 inches); others are 150 millimetres (6 inches) long. Many, but not all kinds, have wings.

Depending on where they live, mantids are different colors and shapes. Many plant-living kinds are green or brown, with patterns like leaves on their wings. Some are gray and bumpy, like the twigs or bark they stand on. Others are pink and shaped like petals, blending with flowers that attract mantid prey.

ON GUARD!

Birds, lizards, skunks, and opossums make meals of mantids — when they can. But it's tough to find these insects. Mantids stand so still and blend in so well that they are hard to see.

A mantid can use the piercing spines on its front legs to attack predators as well as prey.

Once spotted, very few mantids try to fly away. That's something they usually do just to change perches. But some fast-running mantids that live on tree bark may try to out-run danger.

If an attacker grabs its back or middle leg, the mantid might leave the leg behind. It has a special muscle that can let the leg go. Then the mantid can escape. If it is young enough, it may grow a new leg to replace the lost one.

But a mantid usually faces an attacker head on. Rearing up, it stands tall and flares out its wings. To look even more fierce, it exposes any dark, eyelike spots on its body. And as a threat, it holds its front legs high and puffs by jerking its lower body.

The sight may stop an attacker in its tracks. If not, the mantid may lash out with the sharp spines on its front legs. They can damage the eyes of small animals or, at least, scare the animals away.

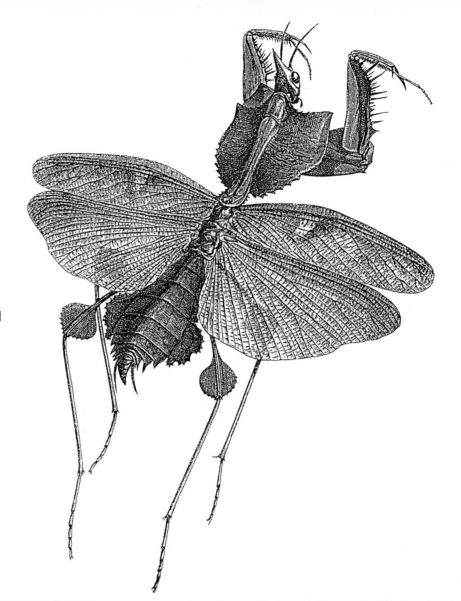

Wings flared, body raised, and spines poised, this mantid is saying, "Stay away!"

MAGIC AND MEDICINE

With front legs raised and folded, mantids look like they are saying their prayers. That's why "praying mantis" is the name given to some kinds. People once believed that mantids came to Earth to show them how to pray — or to help lost shepherds find their way.

Other people were afraid of mantids. They thought it was bad luck to touch one. They believed a mantid's stare could make them sick.

But the Chinese used mantids as medicine. They treated warts, wounds, and illnesses with mantid eggs. Even today, some people try to help themselves feel better by using molted mantid coats.

THE CASE OF THE MOTHER MANTID

When fall comes, the female mantid makes a case to lay her eggs in. She works upside down, using her back end to whip up a foam that comes from her body. Sticking the foam to a plant — sometimes a rock — she builds the case in layers and deposits rows of eggs between them. The foam hardens around each row.

The mantid works for about three hours, building and filling her egg case. Then she leaves. The outside of the case hardens, protecting the eggs and keeping out wind and water. Inside, bubbles of air help shield the eggs from heat and cold.

In spring, the eggs hatch. With legs pressed close to their bodies, mini-mantids worm their way out of the egg case. Up to 400 may hatch out of just one case. But many live only a few hours. Insects, spiders, birds, and lizards snatch them up as meals.

When young mantids stand up, they look just like tiny adults. They grow quickly through the summer, shedding and replacing their coats several times. As they age, many kinds grow darker as well as bigger. But their lives as menacing mantids are short. Most die before winter arrives.

Like other kinds of mantids, this mantid from Asia first emerges from a foamy egg case.

34

EYE WITNESS

It's thrilling to see mantids entering our world. One child made notes as she watched an egg case come alive.

Day 1

- *About 150 mantids wriggled out of the egg case. They looked like worms.*
- *Length: about 15 millimetres [about .6 inches].*
- *Color: light yellow (like tiny cobs of corn) with black eyes.*
- *First they unfolded their legs, then their feelers.*
- *One had its back legs stuck together. I pried them apart with two pins, and it walked right away.*
- *About an hour later, the mantids' color started to change to green. I could see a stripe on their backs. Their eyes got lighter — almost the same color as their heads.*
- *I put a drop of water on my finger. One mantid drank from it.*

Day 2

- *When an aphid came close, a mantid grabbed it fast. One mantid was eating when it caught a second aphid. It held an aphid in each leg.*
- *The mantids cleaned their legs with their mouths. One cleaned its back leg by holding it with its front leg.*

Day 4

- *One more mantid came out of the egg case!*

Day 17

- *Several mantids molted. When they pulled out of their old coats, they were twice as big as when they wriggled out of the egg case.*

CHAPTER 6

As adults, female wasps mostly sip nectar from flowers and sweet sap from trees. But scientists studying yellowjackets — common wasps in cities — discovered they eat nearly everything we do, including hot dogs, ice cream, and pop.

When it comes to getting food for their young, female wasps attack live victims. Some kinds kill and chew up insects. Some paralyze prey and leave it for their young. Others lay eggs in or on a live insect so the young wasps will have food when they hatch.

Wasps are nearly everywhere, but not each of the thousands of kinds is every place. Most have two pairs of wings, but these wings are hooked together so they move as one pair. They are famous for their very slender "wasp waist" — a threadlike part of their body. It allows wasps to move their back ends freely so they can sting or lay eggs in just the right spot. It also allows them to turn around easily in tiny places inside their nests or burrows.

This queen wasp, dining on a fall apple, will become
a fierce hunter to feed her young in spring.

Two bald-faced hornets scrape some bark off a tree to make paper for their nest.

MARVELOUS CHOMPERS

Some wasps, such as yellowjackets and bald-faced hornets, live in colonies with up to 5000 members. Together, they work to feed their young — their larvae — which eat all day.

The wasps hunt for flies and other small insects, grabbing them in flight and crushing them in powerful mouths. Sometimes, the wasps nab prey, such as dragonflies, that are much bigger than themselves. They also snatch up squirming caterpillars. In just an hour, they can capture and carry more than 225 insects to the hungry larvae in their nest.

New colonies start up in spring when the female wasp builds a paper nest full of egg cells. A yellowjacket builds her nest in a sheltered site, such as a hole in the ground or a hollow log. Then she lays her eggs, which take about two weeks to hatch.

All alone, she hunts and feeds the helpless larvae for two weeks, stuffing them with partly chewed insects. Then the larvae spend the next three weeks quiet and still within their cells. They change into adults, and when they emerge, the female — the queen of the colony — has workers to care for the next set of larvae.

All spring and summer, the queen lays eggs. The colony grows — and so does the nest. The workers keep making more layers of cells for more eggs and larvae. Some of them guard the entrance to the nest. Others work to keep a steady temperature inside. They shiver to create heat, and fan their wings or gather drops of water to cool the nest.

By the end of summer, the queen stops laying eggs. A few new queens and some male wasps develop in special cells. The queens mate with the males, then find a winter home, but the rest of the colony dies.

In spring, each queen starts a new nest — and the cycle begins again.

Home for these wasps is a big, weatherproof paper ball.

PAPER MAKERS

Wasp nests are mouth-made. Female wasps scrape bits of wood from trees, sheds — even telephone poles — and chew them up. Their saliva mixes with the wood to form wads of pulp. At the nest site, the wasps spit out the wads and smooth them into tiny strips — paper for building nests.

Each nest is made from hundreds of thousands of tiny mouthfuls of pulp. Many kinds of wasps begin by making a circle of six-sided paper cells. They add rows and rows of cells below that circle and wrap the whole nest in two or three layers of paper. Some nests grow to be 40 centimetres (15.5 inches) across.

Wasp nests are not as fragile as they seem. The paper is thin and light, but it's strong. And air between the layers helps to keep out rain and extreme temperatures.

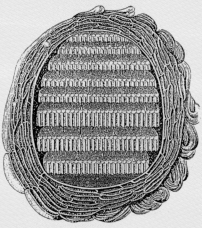

The cells inside a wasp nest form neat rows.

40

POWERFUL PARALYZERS

Many kinds of wasps don't sting to kill; they sting to paralyze. Then they leave their live victim with the eggs they lay. It's food for the larvae that hatch.

One long-legged wasp, the black and yellow mud dauber, makes egg cells from mud under rocks and logs. After spider-hunting, the female uses her head to stuff paralyzed victims into each egg cell. When it holds all it can, she lays an egg in the cell and seals it with mud.

But a smaller wasp, the blue mud dauber, might come along. She never makes her own egg cells. Instead, she breaks into cells made by the black and yellow mud dauber, using water to wet the seals. Yanking out the spiders and eggs that are already there, the blue mud dauber inserts her own. She prefers black widow spiders, laying an egg on the first one she crams into each cell.

One kind of sand wasp uses a tool to close the burrows she digs. After she puts a paralyzed caterpillar inside, she may pick up a small pebble in her mouth and pack dirt down with it. But often, she simply packs dirt with what's handiest — her head.

Before she goes hunting, a wasp such as this one often flies around her burrow and notes what's nearby. That's how she finds her way back again. When scientists experimented by moving little landmarks, the wasp couldn't find the opening to her burrow.

After a successful hunt, a wasp stuffs a stink bug into her burrow.

Potter wasps put paralyzed caterpillars with their wasp eggs into little clay pots. True to their name, the wasps make these pots themselves, molding balls of clay with water

41

and attaching them to plants. Some of the pots look more like a jug with a short neck and wide lip.

Other kinds of wasps put eggs in almost any hole they can find. One jammed some caterpillars and an egg into the mouthpiece of a horn on a country porch. Imagine what happened when someone blew the horn to call the field workers.

One of the most amazing wasps, the giant cicada killer, digs tunnels and makes egg cells underground. Then she hunts a large insect called a cicada to stuff in each of her cells. The wasp stings the cicada until it can't move, then faces the tough job of getting it back to her tunnels.

Hooking her middle legs under the cicada's wings, the wasp often heads for a tree, hauling the insect up to gain height. One even hauled a cicada up the outside of a man's pants. Then, placing her legs on either side of the cicada, the wasp glides toward her burrow with her victim.

TARANTULA HAWKS

A shiny blue wasp with orange wings is out on a hunt. Her feelers sense a tarantula — and the tarantula senses her. Although she is a large wasp, a 30-millimetre-long (1.2-inch-long) tarantula hawk, she's much smaller than the hairy spider.

The tarantula rises on its back legs, its poisonous fangs ready. The wasp curves her back end beneath her body, bringing her stinger forward. Then she attacks. With a powerful bite, she clutches one of the tarantula's legs.

The spider pulls itself free and strikes back, but its fangs fail to pierce the wasp's smooth, curved body. She attacks again, this time stinging the tarantula in a large nerve. The spider falls — limp, but alive.

Pushing and pulling, the wasp moves the tarantula slowly across the sand and down into the spider's own burrow. She clears a spot on its hairy body where she lays and glues her egg. Then the wasp carries sand in her mouth to fill in the burrow, leaving the egg to hatch and a feast for a young wasp to eat.

Wasp larvae have been feeding inside this caterpillar.
After emerging, they formed cocoons, attached to the caterpillar.

INCREDIBLE PARASITES

Wasps that are parasites don't kill or paralyze. They don't even sting. A female parasite wasp usually pierces a young insect or spider, using a thin tube on her back end. She may drink some of the blood that oozes out. Then she lays her eggs inside the victim. For instance, she might lay 15 to 30 eggs in the coat of a large cabbage caterpillar.

When the larvae hatch, they live inside their host. They usually spread themselves out evenly within its body, feeding on blood and fat. At first, the host keeps on living — even growing. But as the wasp larvae get bigger, they start eating other body parts. By the time they burrow out, there's very little left of the host.

Many parasite wasps are extremely small. Some live in the eggs — not the larvae — of other insects. Some are so small they live in the larvae of other kinds of wasps that live in the larvae of still other kinds of wasps that live in caterpillars — all at the same time. One of the smallest parasite wasps is one of the smallest insects in the world: the battledore-wing fairy fly is a speck, just .2 millimetres (.008 inches) long.

One of the largest parasite wasps lays eggs on the giant wood-wasp larva, which tunnels inside trees. The parasite must drill through bark and wood to reach the larva, which she senses with her feelers. For the job, she uses a slender tube as thin as a human hair and 38 millimetres (1.5 inches) long — longer than she is. Twisting and pressing her back end, she forces the sharp cutting tip of the tube right into the wood. Then she squeezes her egg through it to the larva. An "attack wasp" strikes again.

Drilling a hairlike tube through tree bark, a wasp lays an egg in a wood-boring larva inside the tree. The job may take about 30 minutes.

WATER WASPS

Wasps in water? Believe it. One of the fairy flies — among the tiniest of wasps — spends most of its life underwater. Only 1.5 millimetres (.06 inches) long, this fairy fly lays eggs inside the eggs of big diving beetles. The wasp uses its wings to swim as if it were flying, but it moves quite slowly.

Another wasp — smaller but faster — does not "fly" underwater. It "rows" instead, using its legs as oars. Up to 70 of its tiny larvae have emerged from the egg of one diving beetle.

CHAPTER 7

Not every lion is large and furry. Young aphid lions and ant lions are insects that are no bigger than raisins. Yet they are powerful predators — as fierce as the wild cats they are named after.

LURKING LIONS

Stalking prey, the aphid lion creeps toward small insects, such as aphids. It moves with its mouth open and ready, then snaps it shut, piercing the prey. The lion injects juices that dissolve the prey's insides. Then it drains the victim, turning it this way and that to empty it completely. The feeding takes just a minute.

Many kinds of aphid lions wear the leftover outer coats of their prey. Bending their heads back, they use their mouths to hook these coats onto the stiff hairs on their bodies. Some also wear tiny bits of leaves and bark. Looking like piles of garbage helps aphid lions hide from predators, such as birds.

Unlike aphid lions — and most other insects — many kinds of ant lions set traps for prey. With big pincerlike mouths, they dig pits in the sand. They turn as they shovel, making smaller and smaller circles as they dig deeper. In about 30 minutes, they make

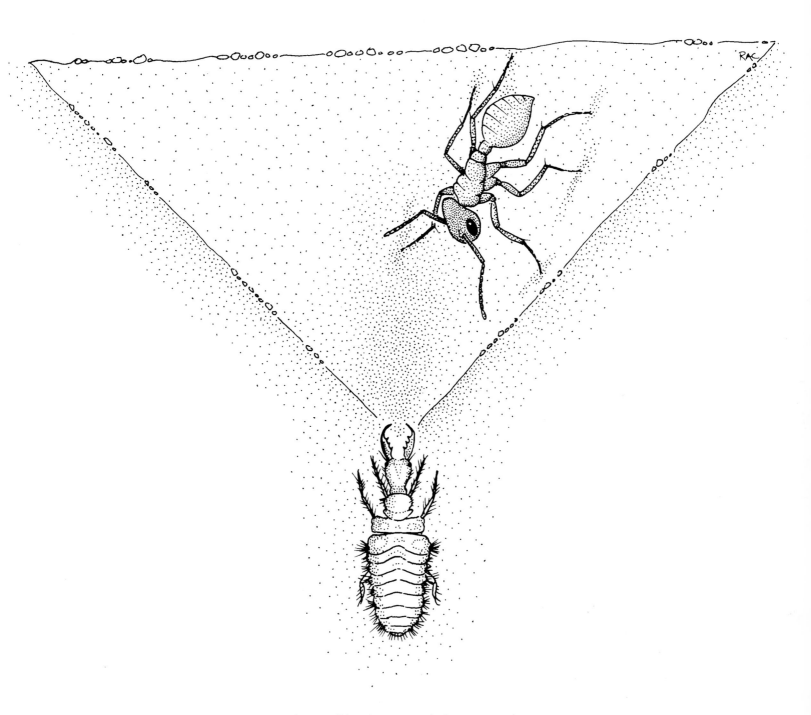

An ant slides down a sand pit — trapped
by the ant lion that waits to grab it.

LIONS OF LACE

Hundreds of kinds of aphid lions and ant lions live around the world. They hatch out of eggs, looking like bristly, oval-shaped monsters. But they mature into lovely wisps on wings.

As adults, both kinds of lions fly with four wings patterned like lace. Some adult aphid lions — called lacewings — are pale green with red-gold eyes; some are brownish with spotted wings.

Grown ant lions look like dragonflies. They are darker and bigger than lacewings. Their bodies may be almost 50 millimetres (2 inches) long — at least twice as long as most lacewings.

smooth, funnel-shaped holes as deep as 5 centimetres (2 inches) and up to 8 centimetres (3 inches) across.

After an ant lion digs a pit, it buries itself in the bottom. Only its grasping mouth sticks out — waiting. An ant crossing the sand may not notice the pit and slip in. If it tries to clamber back up the smooth sides, the ant lion flicks grains of sand to make the slope more slippery.

Down goes the ant into the waiting mouth. The ant lion stabs it and squirts it with juices that dissolve its insides. After feeding, the ant lion tidies its pit, flinging out loose sand and the prey's outer coat. Then it waits again.

This new adult lacewing has just emerged from a cocoon. Its wings have not yet opened.

Caught in a spider's web,
an ant lion comes to the end
of its short adult life.

AS LIONS GROW

The aphid lion grows very fast. When it's 10 to 30 days old, it ties itself to a leaf with silk from its own body. Then it wraps itself completely into a small ball of silk — a cocoon.

About three weeks later, the aphid lion cuts through the top of the cocoon and crawls out. Like magic, it has turned into an adult. Moving slowly, it grabs onto a twig while its body pumps blood to its wings. When they open, the adult aphid lion — a lacewing — flies off.

Like its young, the lacewing hunts small insects, such as aphids. But it no longer hides by covering itself with leftovers. Instead, it has a strong smell — like garlic — which turns some of its predators away. It may also play dead to fool a hungry beetle.

An ant lion must wait longer to become an adult. It can spend up to two years as a pit hunter. In that time, it may dig many pits, making bigger ones as it grows.

One day, it wraps itself up in a silky cocoon under the sand. It may stay there all winter. When it comes out, it is a delicate, winged adult. But unlike the lacewing, the adult ant lion doesn't go hunting. Its life is so short that it doesn't even eat. It simply mates and dies.

BEATING THE BATS

Slow-flying lacewings seem no match for speedy bats. These furry fliers scoop hundreds of insects from the sky each night.

But the lacewing uses a special trick to escape. In each of its front wings, there's an ear that can hear the high-pitched cries of a bat. As the hungry bat flies near, the lacewing closes its wings and plunges. It dives at speeds of about 2 metres (6.5 feet) per second.

The bat follows. But just as it gets close, the lacewing performs a quick flip, changing the path of its power dive. The shrieking bat aims for the spot where the lacewing would have gone before the flip — and misses its target.

LAYING LION EGGS

After mating, the female ant lion lays her eggs in the sand. That way, the young can start digging their pits soon after they hatch.

But the female lacewing makes "stems" to hold her eggs. With her back end, she dabs a leaf or twig with a thick, sticky liquid. Then she stretches it upward — like a strand of gum — to form a threadlike stem. On top, she lays an egg.

Over and over, the lacewing dabs and stretches, dabs and stretches. In four weeks, she lays hundreds of eggs-on-a-stem. Each one hatches in about 10 days — unless ants crawl up the stems and attack the eggs.

Each young aphid lion cuts a lid in the top of the egg. Once out, the lion dries a bit, then slips down the stem to start hunting for food. If the lacewing hadn't set her eggs on stems, the first lurking lion to hatch could have easily reached them — and might have eaten them all.

After mating, male and female lacewings go their separate ways.

DOODLEBUGS DOODLE

Ever doodle in the sand? Ant lions do. As soon as they hatch, they start looking for places to dig their ant-catching pits. They wander across the sand, always moving backward. Thick, bristly body hair that points frontward forces them to walk that way. As they travel, the ant lions make wiggly lines that look like doodles in the sand. That's why some people call them "doodlebugs."

51

Beetles live everywhere, and the number of kinds is amazing! Scientists have found about 300 000 different ones, but there are likely even more.

EVER-EATING BEETLES

Some are as short as .25 millimetres (.01 inches); others are over 200 millimetres (8 inches) long. But their hard, thick wing covers make them easy to spot as beetles.

Not only are there many kinds of beetles, there are billions and billions of individuals. Many chomp on plants to feed, but many go hunting for meat. So when beetles get hungry, other insects better watch out.

Among the petals of a dandelion, the ladybug
may snack on pollen while waiting for aphids.

LADY LUCK, LADY LOVE

Ladybugs are the world's most popular beetles. For centuries, people thought ladybugs brought good luck, good love, good weather, good crops — even good health. Some thought they cured illnesses, such as measles; others thought that crushed ladybugs, stuffed in cavities, cured toothaches.

Before the 1500s, European farmers believed the ladybug could do so many miracles that they dedicated it to "Our Lady, the Virgin Mary." That's why "lady" is part of many of the beetle's names, including ladybird, lady beetle, ladyfly, and ladycow.

RAVENOUS WOLVES

Eating is what a young ladybug — the larva — does best. This beetle spends most of its young life wolfing down insects, such as aphids. Even when it has gorged itself, it keeps on eating, taking just a bite out of each aphid it grabs. In fact, it may eat up to 500 aphids a day. No wonder it is called an aphid wolf.

But this wolf is one that doesn't have to move fast on the hunt. Its prey barely moves at all. And the aphid wolf doesn't need big, wolflike jaws either. It just nabs aphids and other soft insects it finds on plants, and chews them up.

If prey is scarce, the aphid wolf still manages fine. It can go many days without eating. It can also switch to flowers, feeding on their pollen and nectar.

As the spiny, wrinkled wolf grows, it sheds its coat three times in about five weeks. Then it glues its back end to a leaf or twig and waits quietly while its body changes. Ten days later, it bursts out of its coat as an adult ladybug up to 18 millimetres (.7 inches) long.

During the hours that follow hatching, all the spots the ladybug will ever have appear on its wing covers. The number depends on which of the thousands of kinds of ladybugs it is.

As an adult, the ladybug still eats a lot. Its mouth is wide and its teeth are jagged. It searches for aphids along leaves — even underneath them. Holding on is no problem for an upside-down ladybug because it has sticky pads at the ends of its legs.

If danger threatens, the ladybug may lie on its back and play dead. It may ooze a bad-smelling, bad-tasting liquid from its leg joints. As an aphid wolf, it also puts out this sticky yellow blood, which makes predators back away.

After eating for a few weeks, adult ladybugs head for a place to sleep through the winter. Some go alone; others gather together. Hundreds of millions of ladybugs — enough to bury four football fields — have met at a single spot. Together, they are warmer and safer.

Hunting food keeps ladybugs busy — whether they are adults (those on the left) or larvae (those on the right).

RACING TIGERS

With wide, watchful eyes, tigers stalk their prey, sometimes running, then pouncing. But for their size, tiger beetles are faster, more ferocious hunters.

When their big, bulging eyes spot prey, tiger beetles charge. Their long legs cover up to 60 centimetres (2 feet) a second — a speed record for land beetles. At that rate, if a 12-millimetre (.5-inch) tiger beetle grew as big as a horse, it could run 400 kilometres (250 miles) an hour.

Quick feet make these tigers hard to catch. They run to escape robber flies and other predators. They are also quick to spread their wings and take to the air when disturbed.

On open fields and sandy shores, tiger beetles hunt many kinds of insects, attacking

Tiger beetles come in many colors and lengths, up to 40 millimetres (1.5 inches). This little green hunter is about to race across the sand.

them with long, sharp mouthparts. They keep especially busy on days that are hot and sunny. Clouds and shadows seem to slow them down.

Even a young tiger beetle — looking like a bumpy, white caterpillar — has strong, sharp mouthparts. It pokes these out of the entrance to its narrow burrow in the ground. Hooks and spines on its back anchor its body to the burrow's sides.

Ever ready, the tiger snatches passing insects and pulls them into the burrow. Even insects, such as dragonflies, that are several times bigger than the beetle become meals.

As the young tiger grows, it makes its burrow larger, shoveling out dirt with its mouth. Before it turns into an adult, it digs a special room where its body quietly changes.

One day, the tiger emerges as a handsome, grown beetle that may be green, gold, or other shades, depending on its kind.

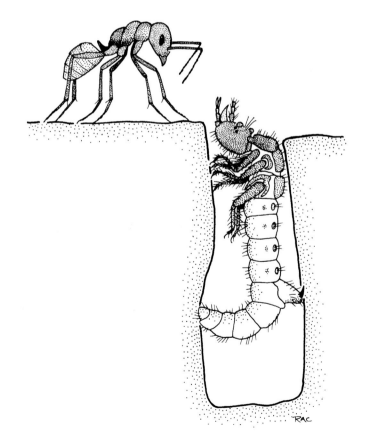

Peering out from its sand tunnel, a young tiger beetle is ready to snatch a passing ant.

WATER TIGERS

Creeping among underwater plants, there's a long, slender tiger with a huge appetite. It's a water tiger — the larva of the diving beetle. With strong, sharp mouthparts, this tiger seizes prey and injects fluids that paralyze and digest.

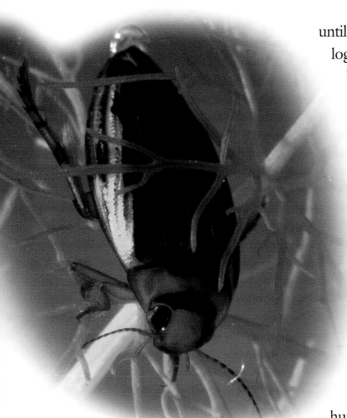

A grown diving beetle prepares to hunt. Note the bubble that has formed as the beetle takes in air.

The water tiger eats and grows, and grows and eats until it is ready to crawl onto shore. In the moist soil under logs and stones, its body gradually changes, and the larva becomes an adult. But it still has a huge appetite.

The grown diving beetle heads back to the water to hunt. It's well built for swimming. Oval-shaped and smooth, its body slips easily through the water. Its flattened back legs, fringed with long hair, work like oars to power the beetle. As it swims, its middle legs steer, leaving its front legs free to capture prey.

Up to 70 millimetres (2.75 inches) long, the diving beetle attacks all kinds of water insects and other small animals, such as snails, fish, toads, and salamanders. Sometimes it also attacks other diving beetles. Whatever the prey, the beetle chews it up in chunks.

Now and then, the diving beetle interrupts its hunting and swims to the surface for air. It stores a large supply under its wing covers for breathing underwater. It also breathes air trapped on plants beneath the water's surface.

Many diving beetles live for several years, flying from pond to pond. Most kinds are good fliers, but sometimes they crash-land. If they mistake the roof of a shiny car for a patch of shiny water, they dive straight into it. And sometimes they smash into windows because they are attracted to lights.

GAPING JAWS

Almost anything small and moving can bring a ground beetle running. It dashes out on strong legs with its mouth held wide open. But it has to get very close before its feelers sense whether or not the object is prey.

These special ground beetles, called bombardier beetles, defend themselves by shooting stinging gas.

Some kinds of ground beetles seize prey and tear it apart with scissorlike movements. Others inject fluid to digest the prey before feasting on it.

As you might guess, many ground beetles hunt on the ground. But some kinds hunt underground, and others hunt well above ground — in trees. A few even live by the sea where the tide sometimes covers them. They likely survive by breathing little pockets of air trapped in cracks in rocks.

Both adult and young ground beetles are hungry hunters. Up to about 35 millimetres (1.4 inches) long, they feed on snails, slugs, moths, and especially caterpillars. A young beetle, or larva, can gobble up more than 50 caterpillars in a couple of weeks. An adult eats hundreds in its life of two to four years. One researcher found a ground beetle that had stuffed itself so full its stiff wing covers bulged up.

Nighttime is when most ground beetles go hunting. Although some kinds are bright colors, most are black or brown, blending with the darkness. During the days, they crawl under dead leaves and rocks, where they rest. Then these ever-eating beetles can go chomping again.

BEETLE BOMBER

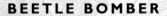

A little gray-and-orange insect called a bombardier beetle defends itself like nothing else. When a predator threatens, chemicals stored in the beetle's body flow together and explode. The mixture of those chemicals gets so hot it boils.

The beetle swivels its back end to take careful aim, and POP! It fires out a smelly, burning, smokelike gas. Again and again the beetle fires, until the heat, smell, and noise turn away insects, ants, spiders, mice, frogs, and birds. Then the beetle bomber scoots off.

INDEX

PHOTOGRAPH AND ILLUSTRATION CREDITS

Photographs

All photographs from the Royal British Columbia Museum except for those by Steve Marshall on pages 5, 31, 32, 37, 38, 41, 44, 53, 58, 59; and by the Canadian Forest Service on page 43.

Illustrations

Illustrations on pages 20, 24, 47, 57 by Rob Cannings/Royal British Columbia Museum.

Illustrations on pages 6, 14, 33, 40 (left) from *A Pictorial Archive from Nineteenth Century Sources.*

Illustration on page 40 (right) from *The Insect World* by Louis Figuer. London: Cassell, Peter and Galpin.

Illustration on page 55 from *The Transformations of Insects* by P. Martin Duncan. London: Cassell, Peter and Galpin.

ACKNOWLEDGMENTS

I am especially grateful to Rob Cannings of the Royal British Columbia Museum for reviewing my manuscript and producing some of the photos and drawings for this book; to Brent Cooke of the Royal British Columbia Museum for supporting this project and contributing some of the photos; to Steve Marshall of the University of Guelph for providing photos; to my daughter, Carolyn, for sharing her enthusiasm for insects and contributing her observations of mantids hatching and developing; and to my husband, Wayne, for joining me on many trips to fields, forests, and ponds to glimpse the incredible world of insects.

The author greatly appreciates the assistance of the British Columbia Arts Council.

The Royal British Columbia Museum gratefully acknowledges the work of George Doerksen.